XTREME AIRCRAFT

UNMANNED AIRCRAFT

A&D Xtreme
BOLD HI-LO NONFICTION
An imprint of Abdo Publishing
abdobooks.com

S.L. HAMILTON

TAKE IT TO THE XTREME!

GET READY FOR AN XTREME ADVENTURE!
THE PAGES OF THIS BOOK WILL TAKE YOU INTO
THE THRILLING WORLD OF UNMANNED AIRCRAFT.
WHEN YOU HAVE FINISHED READING THIS BOOK, TAKE THE
XTREME CHALLENGE ON PAGE 45 ABOUT WHAT YOU'VE LEARNED!

ABDOBOOKS.COM

Published by Abdo Publishing, a division of ABDO, PO Box 398166, Minneapolis, Minnesota 55439.

Printed in the United States of America, North Mankato, MN.

102021

012022

THIS BOOK CONTAINS RECYCLED MATERIALS

Editor: John Hamilton

Copy Editor: Tamara L. Britton

Graphic Design: Sue Hamilton

Cover Design: Laura Graphenteen

Cover Photo: Shutterstock

Interior Photos: Amazon-pg 44; AP-pgs 12-13, 20 & 24-25; BAE Systems-pgs 6-7; DJI-pgs 11, 14-15, & 32-33; Getty/i-Stock-pgs 22-23; Intel-pg 25 (top inset); Monash University-pg 8 (top); NASA-pgs 28-29 & 30-31; Shutterstock-pgs 1, 10, 16-17, 18-19, 20 (inset), 34-35 & 42-43; Teledyne FLIR-pg 36 (inset bottom); US Air Force-pgs 40-41; US Army-pgs 36-37 & 38-39; US Coast Guard-pgs 4-5; US Marines-pgs 37 (top inset); Wikimedia-pgs 8 (bottom) & 9; XM2 Pursuit-pgs 26-27; Yuneec-pg 21.

LIBRARY OF CONGRESS CONTROL NUMBER: 2021943582

PUBLISHER'S CATALOGING-IN-PUBLICATION DATA

Names: Hamilton, S.L., author.

Title: Unmanned aircraft / by S.L. Hamilton

Description: Minneapolis, Minnesota : Abdo Publishing, 2022 | Series: Xtreme aircraft | Includes online resources and index.

Identifiers: ISBN 9781532197376 (lib. bdg.) | ISBN 9781098219574 (ebook)

Subjects: LCSH: Aviation--Juvenile literature. | UAVs (Unmanned aerial vehicles)--Juvenile literature. | Drone aircraft--Juvenile literature. | Reconnaissance aircraft--Juvenile literature.

Classification: DDC 629.1333--dc23

TABLE OF CONTENTS

UNMANNED AIRCRAFT 4

DRONE HISTORY 6

RECREATIONAL DRONES 10

COMMERCIAL DRONES 14

ENTERTAINMENT DRONES 24

SEVERE WEATHER DRONES 28

SEARCH & RESCUE DRONES 32

MILITARY DRONES 36

BECOMING A DRONE PILOT 42

THE FUTURE 44

XTREME CHALLENGE 45

GLOSSARY 46

ONLINE RESOURCES 47

INDEX 48

CHAPTER 1

UNMANNED AIRCRAFT

Unmanned aircraft are flown by pilots on the ground or are programmed to fly a specific route. They are called drones or unmanned aerial vehicles (UAVs). These aircraft are flown for fun, but also perform important jobs in the military and many other industries.

XTREME FACT

In the United States, all drones over .55 lb (250 gm) must be registered with the Federal Aviation Administration (FAA).

A MQ-8 Fire Scout lands on a US Coast Guard ship. With sea-scanning radar, the UAV can help with search and rescue operations.

CHAPTER 2

DRONE HISTORY

Military training inspired the first unmanned aircraft. In the 1930s, United Kingdom (UK) leaders realized pilots and soldiers needed a lot of practice to accurately shoot down enemy aircraft. In 1935, the De Havilland Aircraft Company created the DH82 Queen Bee as a target for anti-aircraft gunnery training. The full-size aircraft was radio-controlled by an operator on the ground. More than 400 Queen Bees were built.

The DH82 Queen Bee used a catapault launch.

If DH82 Queen Bees were not destroyed in target practice, they were brought down and used again. Some had wheels, while others had floats for water landings.

XTREME FACT

Honoring the UK's Queen Bee, Commander Delmer Fahrney of the US Naval Research Laboratory came up with the term "drone" (a worker bee) for the US Navy's own unmanned target aircraft. The name "drone" stuck.

Reginald Denny sits by a RP-1. The US Navy referred to it as a TDD (Target Drone Denny) or a "Tiddledy-dee."

Actor Reginald Denny began the Radioplane Company after becoming fascinated with a neighbor's radio-controlled (RC) toy plane. Radioplane One (RP-1) was manufactured, demonstrated, and sold to the US Army and Navy as a target drone in 1938. These Radioplanes became the United States' first mass-produced UAVs.

During World War II, the US Army used nearly 15,000 Radioplane target drones that they called OQ-2s, OQ-3s, and other models.

XTREME FACT

Norma Jean Dougherty was photographed assembling Radioplane target drones in 1944. The photos helped launch her successful acting career as Marilyn Monroe.

Norma Jean Dougherty, holds an OQ-2 propeller at a Radioplane factory in Burbank, California.

CHAPTER 3

RECREATIONAL DRONES

Drones used for fun usually have about 30 minutes of flight time.

Recreational drones flown for fun are typically lightweight, but strong enough to withstand a few crashes. They can fly in gentle winds, but gusts may ground the small drones. Many come with tracking ability to follow moving objects, as well as **obstacle avoidance**.

Drones such as the DJI Air 2S have still and video cameras that take incredible shots and footage in a variety of cool, humid, and hot locations.

FPV (First Person View) racing involves operators wearing goggles and racing their drones through a set course. The race is live-streamed from each drone's camera so an audience can watch. Many racers build their own drones.

FPV goggles make operators feel like they are in their drone's cockpit.

XTREME FACT

Drone racing began in abandoned buildings in the early 2000s. By 2015, the first National Drone Racing Championship was held in California.

CHAPTER 4

COMMERCIAL DRONES

Commercial drones are UAVs flown for work. They are used in many industries, including businesses that need views of remote, dangerous, or high places.

Inspection and construction industries use drones such as DJI's Matrice 300 RTK to look over everything from pipelines and electrical towers to buildings and bridges. Commercial drones usually fly for 55 minutes and can take video and still photos.

A DJI Matrice 300 RTK is used to examine a tall oil structure.

Drones are flown to inspect indoor and underground areas that may be dark, smoky, or dangerous. These sturdy UAVs carry bright lights and high-quality cameras as they fly into mines, tunnels, sewers, smokestacks, and storage tanks.

Drones can be used to inspect and map mines.

An agriculture drone is used to scan fields to see where the soil is wet and where it needs watering.

Farmers use drones to help them determine which fields are growing well and which ones need help. Some drones carry tanks that can be filled with liquid. They can spray fertilizer, as well as bug or weed killer on crops.

Drones with 2.6-gallon (10-L) tanks spray fertilizer evenly across corn crops. Sensors keep the UAVs flying level with the crops below them.

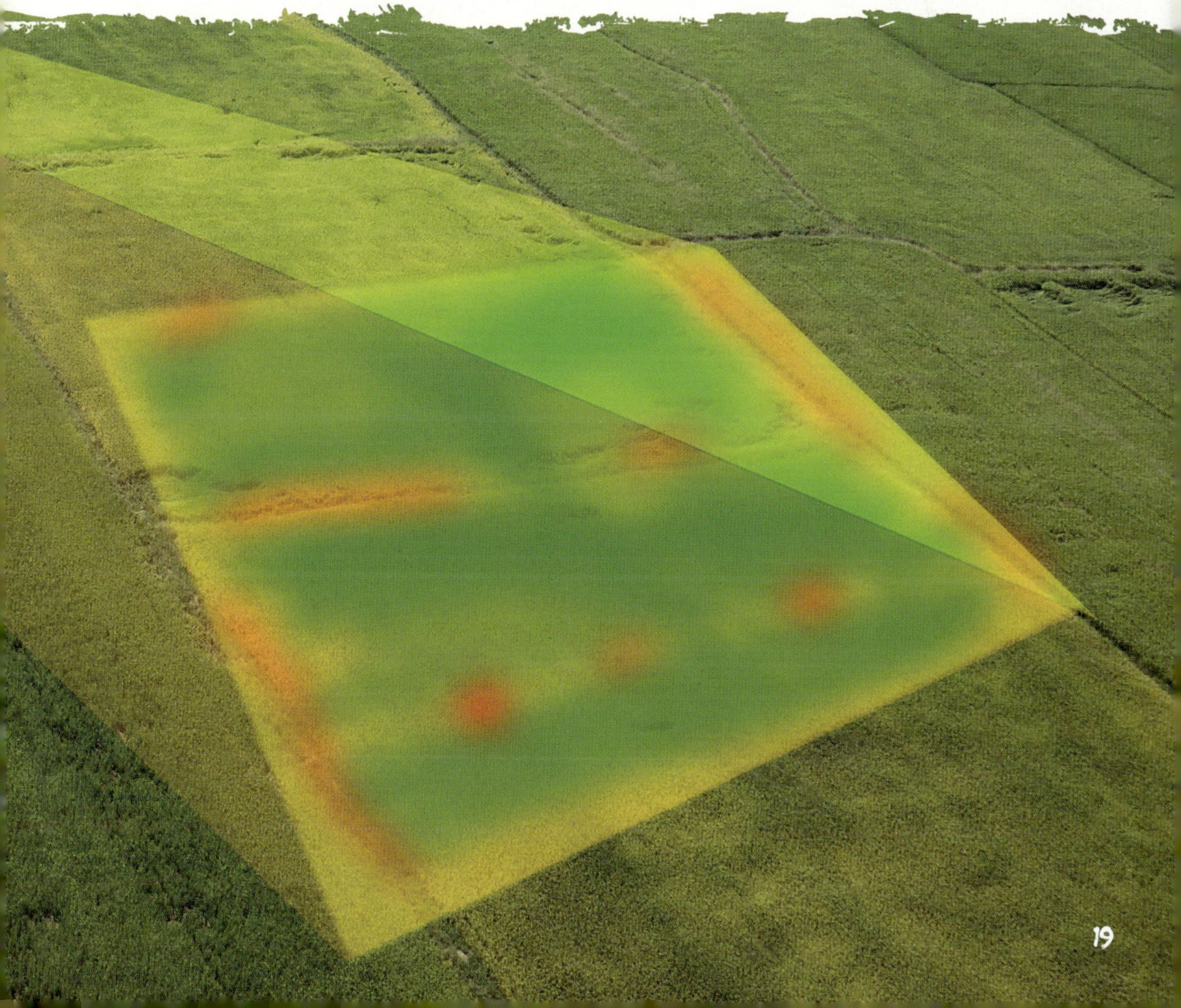

One of the top commercial uses for UAVs is in real estate. Drones such as DJI's Mavic Mini are used to photograph or video the outside and inside of homes and businesses that are for sale.

A drone photo of a home for sale.

The Mavic Mini has a range of up to 2.5 miles (4 km). It is small and easily transportable to real estate locations.

Yuneec H520's software creates a map of a desert area.

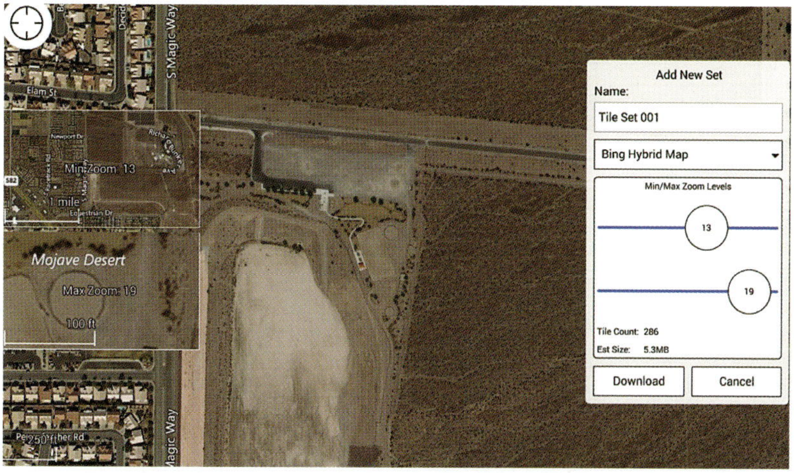

Commercial drones are also used for mapmaking. The Yuneec H520 is a **hexacopter** with software that lets the UAV map an area using overlapping image tiles or a single large image.

A drone tracks an elephant in Botswana, Africa. In the US, a special permit is needed to fly in national parks or wilderness areas.

Wildlife management organizations use UAVs to count and track wildlife. Some drones shoot darts that put large animals to sleep for wellness checks or inject them with medicine.

In the United States, drones are mostly used for counting shorebirds and seals. **FAA** rules state that non-recreational drone pilots may only fly during the day and must stay below 400 ft (122 m). Pilots must always be able to see their UAVs.

CHAPTER 5

ENTERTAINMENT DRONES

Drones are used in the entertainment industry, both for filmmaking and as part of light-formations shows. Multiple UAVs, often combined with lasers or fireworks, can be programmed to create exciting moving pictures in the sky.

A total of 1,824 Intel Shooting Star drones flew in formation to create a spinning globe in the sky at the Summer Olympics in Tokyo, Japan, in 2021.

The entertainment industry uses special heavy-duty drones for aerial shots. UAVs are cheaper to use than helicopters and can be flown in tight areas. XM2's Sierra and Tango II drones can carry up to 66 pounds (30 kg) of camera gear and batteries.

An XM2's Sierra drone is loaded with 8 batteries and an ARRI film camera. A Sierra can fly for 10-12 minutes.

XTREME FACT

Drone pilots wearing safety gear may be secured onto a flatbed truck and driven close to the actors and stunts to film the action.

XM2's Tango II drone is used to film the movie *F9: The Fast Saga*.

CHAPTER 6

SEVERE WEATHER DRONES

Global Hawks were originally built by Northrop Grumman for the US Air Force. The UAVs are 44 feet (13 m) long and travel at a cruising speed of 348 mph (560 kph).

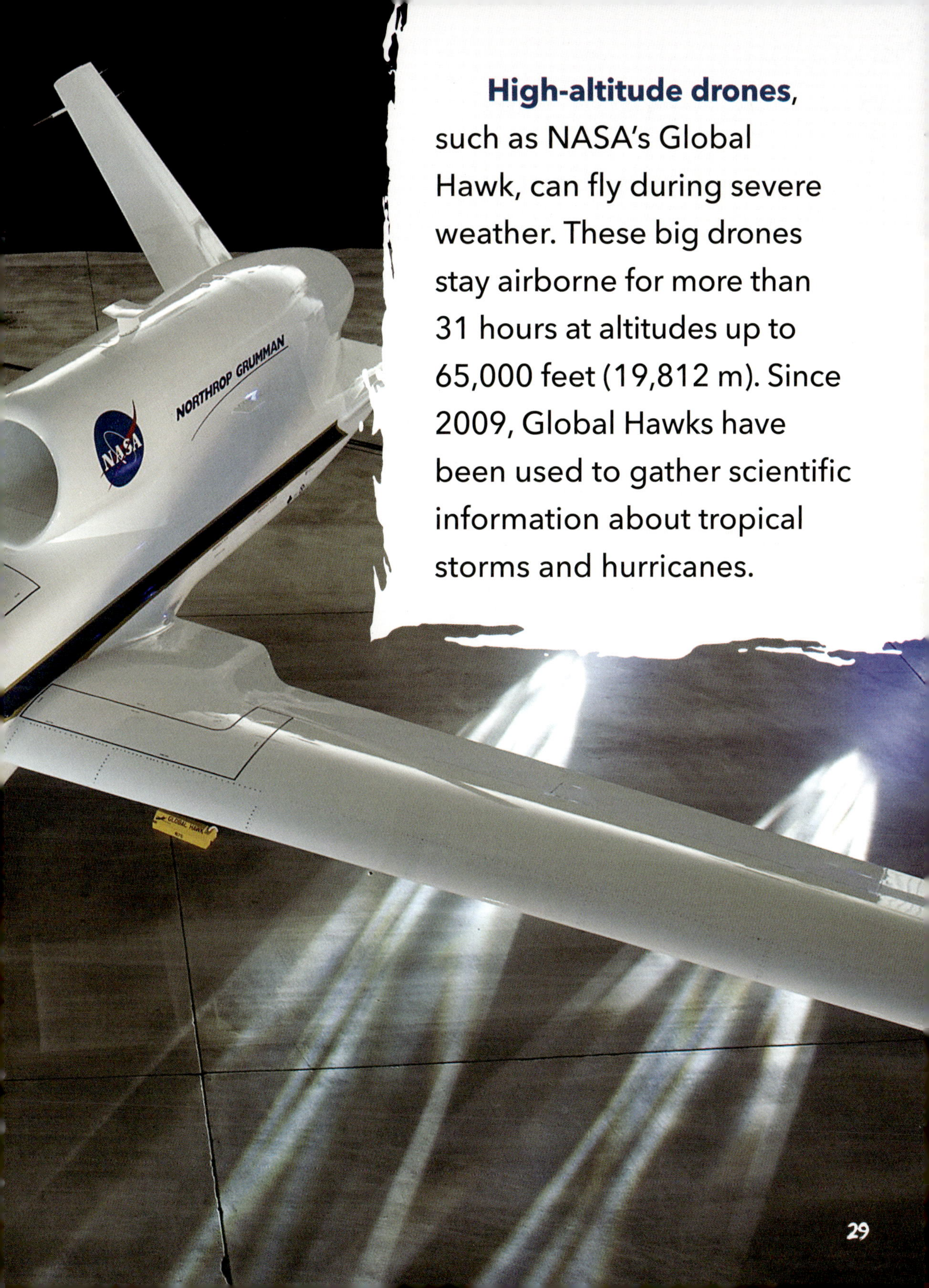

High-altitude drones, such as NASA's Global Hawk, can fly during severe weather. These big drones stay airborne for more than 31 hours at altitudes up to 65,000 feet (19,812 m). Since 2009, Global Hawks have been used to gather scientific information about tropical storms and hurricanes.

Global Hawks are part of NASA's Hurricane and Severe Storm Sentinel (HS3) mission. The HS3 UAVs are loaded with scientific equipment designed to research the formation of hurricanes and how these storms change in intensity. The information gathered will help warn people of approaching severe storms.

A Global Hawk returns to NASA after gathering data on Tropical Storm Edouard over the North Atlantic Ocean in 2014.

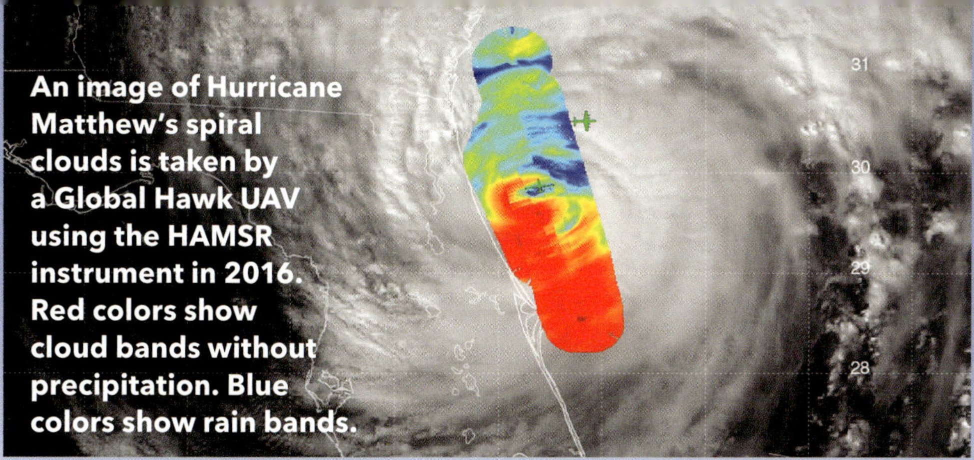

An image of Hurricane Matthew's spiral clouds is taken by a Global Hawk UAV using the HAMSR instrument in 2016. Red colors show cloud bands without precipitation. Blue colors show rain bands.

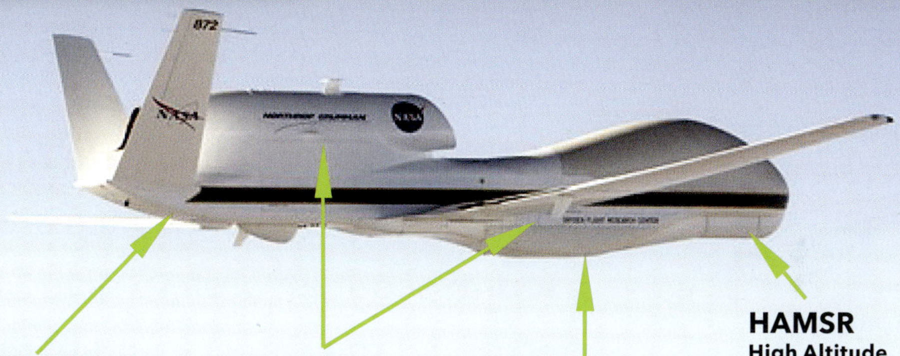

Dropsonde
A single-use device that measures temperature, humidity, pressure and winds as it falls to the sea surface.

LIP
Lightning Instrument Packages track and document lightning as hurricanes develop and intensify.

HIWRAP
High-Altitude Imaging Wind & Rain Airborne Profiler measures horizontal and ocean surface winds.

HAMSR
High Altitude MMIC Sounding Radiometer measures water vapor, cloud liquid and ice distribution, and more.

CHAPTER 7

SEARCH & RESCUE DRONES

The DJI Matrice 200 can fly 35 minutes in cold temperatures.

Search and rescue (SAR) drones are vital tools to help locate missing and injured people. Drones are also used to scout an area to give SAR personnel the safest route to reach someone in dangerous terrain.

SAR drones such as DJI's Matrice 200 have both video and **thermal cameras**. Even in the cold and dark, drones are able to search wide areas of land.

The Matrice 200 has a loudspeaker so pilots can talk and listen to lost or injured people. SAR personnel can then bring any needed rescue items.

A drone's thermal camera displays the locations of people on the ground.

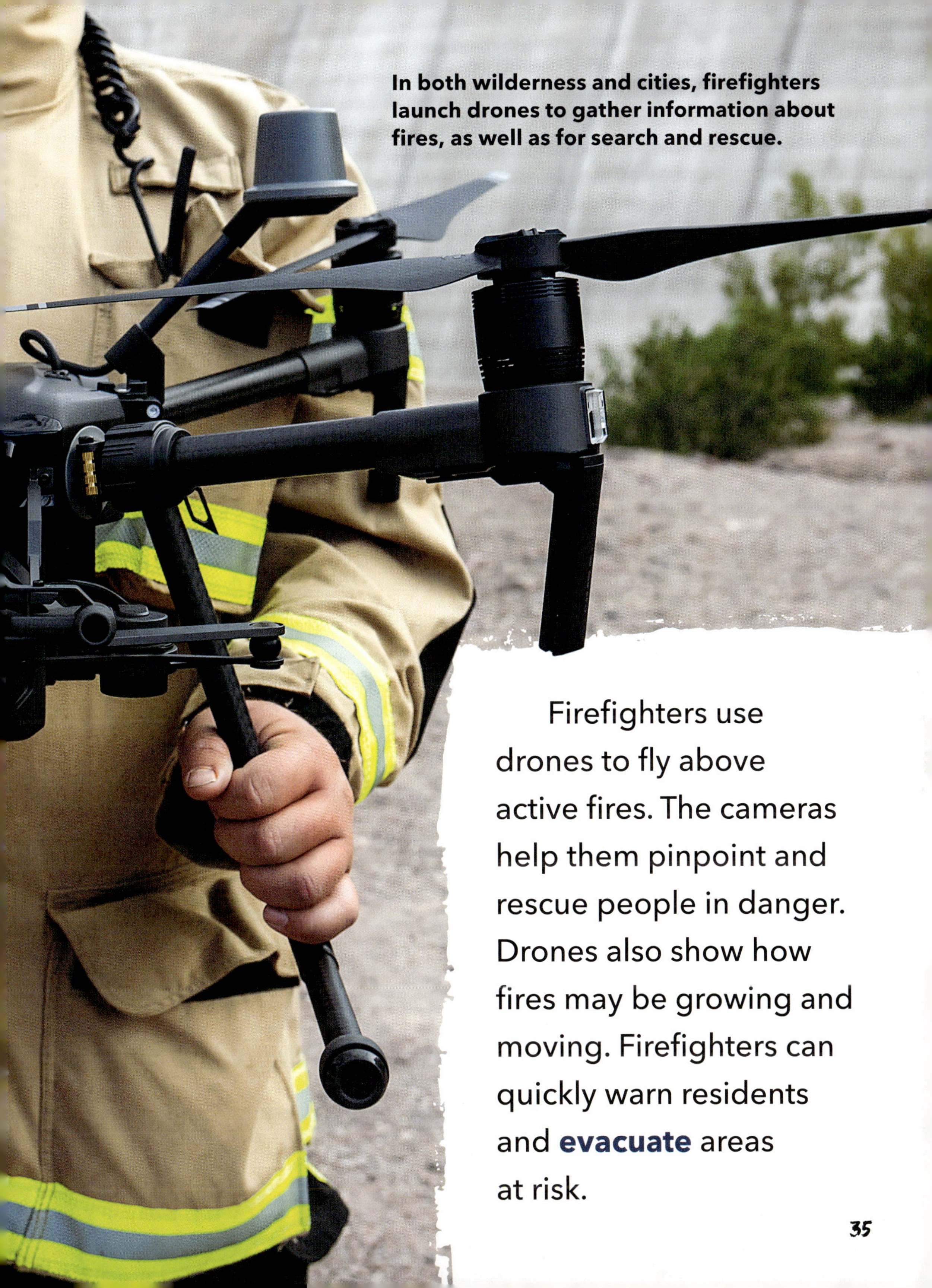

In both wilderness and cities, firefighters launch drones to gather information about fires, as well as for search and rescue.

Firefighters use drones to fly above active fires. The cameras help them pinpoint and rescue people in danger. Drones also show how fires may be growing and moving. Firefighters can quickly warn residents and **evacuate** areas at risk.

CHAPTER 8

MILITARY DRONES

Drones have been used by the military for decades. FLIR's Black Hornet VRS (vehicle **reconnaissance** system) takes spy drones to a tiny size. The finger-length, .56-oz (16-gram) drone can be flown outdoors, as well as inside buildings. It may be used for intelligence gathering, **surveillance**, and reconnaissance.

A Black Hornet is launched from a soldier's hand. It flies nearly silently for about 25 minutes.

A Black Hornet's screen shows the area.

The Black Hornet (shown actual size) is 4 inches (10 cm) long.

A soldier holds a Black Hornet's controller in his hand. The display screen is attached to his vest.

An RQ-11 Raven has a wingspan of 4.6 ft (1.4 m) and weighs 4.2 lbs (1.9 kg).

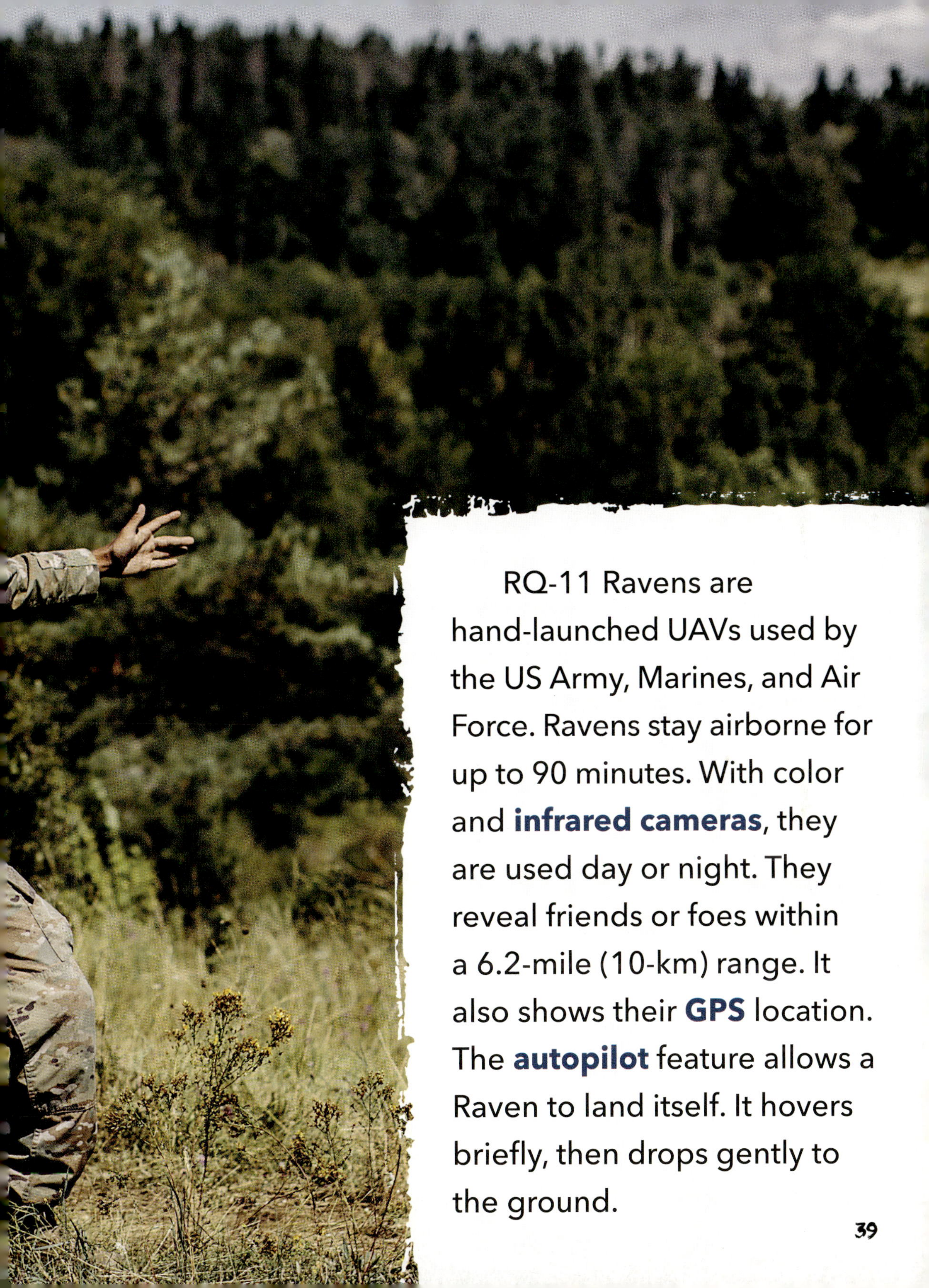

RQ-11 Ravens are hand-launched UAVs used by the US Army, Marines, and Air Force. Ravens stay airborne for up to 90 minutes. With color and **infrared cameras**, they are used day or night. They reveal friends or foes within a 6.2-mile (10-km) range. It also shows their **GPS** location. The **autopilot** feature allows a Raven to land itself. It hovers briefly, then drops gently to the ground.

The MQ-9 Reaper and MQ-1B Predator are medium-altitude, long-range UAVs used on many military missions. They may carry combat weapons for precision strikes.

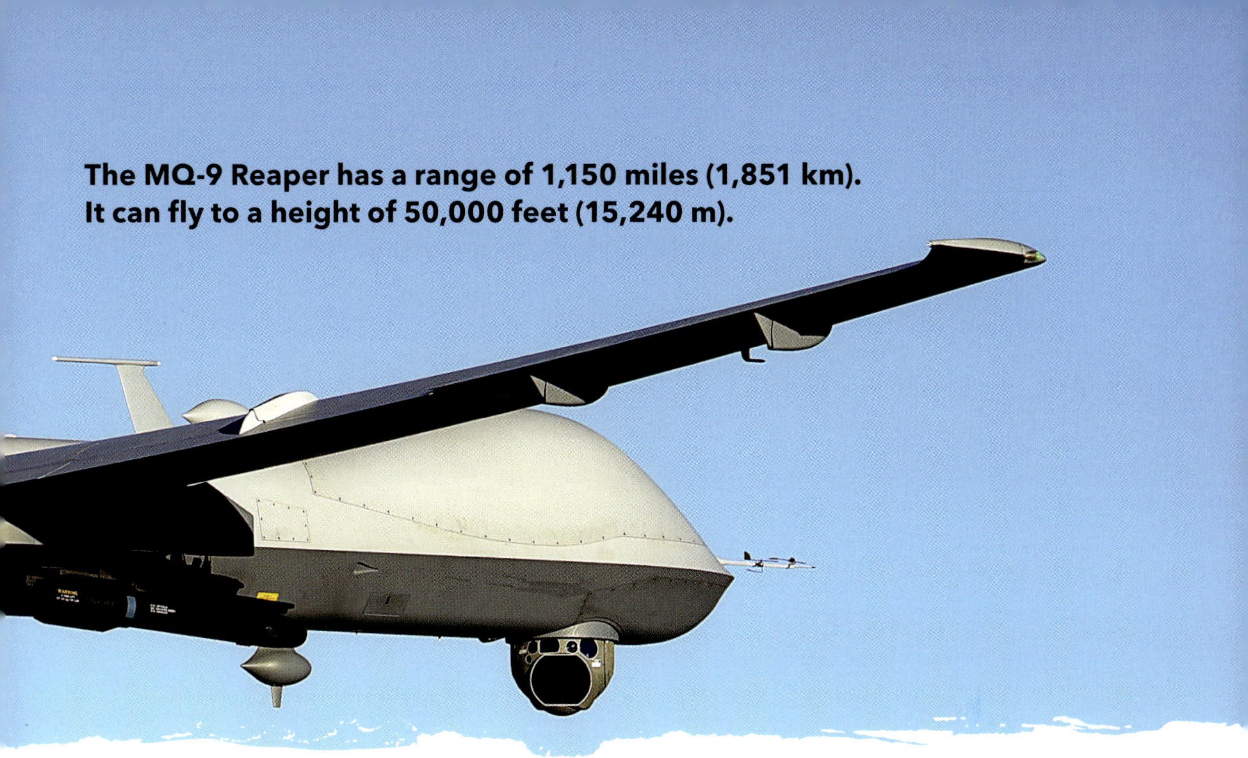
The MQ-9 Reaper has a range of 1,150 miles (1,851 km). It can fly to a height of 50,000 feet (15,240 m).

They carry sensors for **surveillance** and **reconnaissance**. The drones have ground-based crews, including a pilot and 1-2 aircrew members to operate sensors and guide weapons.

The MQ-1B Predator has a range of 770 miles (1,239 km). It can fly to a height of 25,000 feet (7,620 m).

CHAPTER 9

BECOMING A DRONE PILOT

To fly a commercial drone in the United States, a person needs a Remote Pilot Certificate from the **FAA**. To get it, a person must pass an exam called the Unmanned Aircraft General-Small (UAG). The UAG test makes sure people know where and when they can fly their drones. There are online, in-person, and military schools for drone training.

XTREME FACT

Recreational drone flyers do not have to be licensed. They do have to follow the FAA's rules for drone use.

A person must be at least 16 years old to get a Remote Pilot Certificate.

CHAPTER 10

THE FUTURE

Drones of the future will be able to fly longer, faster, and carry greater loads. UAVs will come in many sizes, from tiny eyes in the sky to drones big enough to be used as package delivery, or even individual air taxis. Recreational drones will be faster, more **maneuverable**, and fun to fly and race.

Amazon Prime Air is a planned drone delivery system. It would take a package up to 5 lbs (2.3 kg) less than 30 minutes to arrive.

XTREME CHALLENGE

TAKE THE QUIZ BELOW AND PUT WHAT YOU'VE LEARNED TO THE TEST!

1) What does UAV stand for?

2) What were the first unmanned planes used for in the military?

3) Recreational drone racers use FPV goggles to feel like they are in their drone's cockpit. What does FPV stand for?

4) Agriculture drones often carry tanks of liquid. What might they be spraying on crops?

5) What large, high-altitude UAV is used by NASA to gather information on tropical storms and hurricanes?

6) What are military UAVs used for?

7) To fly a commercial drone in the US, a person must get what kind of certificate from the FAA? Do recreational drone flyers need this, too?

GLOSSARY

autopilot – A device that automatically flies an aircraft, or other vehicle, on a set course, without a human in control.

evacuate – To remove people, animals, and other living things from a dangerous place to a safe area.

FAA – Federal Aviation Administration. The FAA is part of the US Department of Transportation. It is charged with maintaining the safety of civil (non-military) aviation. The FAA enforces the laws and makes sure minimum standards are met in the manufacture, operation, and maintenance of aircraft flown in the United States.

GPS – Global Positioning System. Orbiting satellites transmit information to GPS receivers on Earth. Using this information, receivers can calculate location, speed, and direction with great accuracy.

hexacopter – A helicopter-like drone with six rotors.

high-altitude drones – A type of unmanned aircraft that flies at heights of 60,000 ft. (18,288 m) or more. Many also remain airborne for extremely long periods of time. Some high-altitude drones can remain aloft for up to 35 hours.

infrared camera – A type of camera that forms an image by measuring infrared radiation. Infrared refers to a certain range of electromagnetic radiation. Visible infrared waves can show variations in temperature, with hotter objects a different color than cooler things.

maneuverable – Able to move quickly and easily.

reconnaissance – To explore or scout an area.

surveillance – To closely watch suspected spies, criminals, or enemies.

thermal camera – A camera with sensors that detect different amounts of infrared energy (heat). It converts the temperature differences into a visual image.

obstacle avoidance – Sensors in a drone's programming that allow it to go around objects in the way of its flight path.

ONLINE RESOURCES

Booklinks
NONFICTION NETWORK
FREE! ONLINE NONFICTION RESOURCES

To learn more about unmanned aircraft, please visit **abdobooklinks.com** or scan this QR code. These links are routinely monitored and updated to provide the most current information available.

INDEX

A
Africa, 22
Air Force, US, 28, 39
Air 2S (DJI), 11
Amazon Prime Air, 44
Army, US, 8, 39
ARRI, 26

B
Black Hornet VRS, 36, 37
Botswana, Africa, 22
Burbank, CA, 9

C
California, 9, 13
Coast Guard, US, 5

D
De Havilland Aircraft Company, 6
Denny, Reginald, 8
DH82 Queen Bee, 6, 7
DJI, 11, 15, 20, 32, 33
Dougherty, Norma Jean, 9
Dropsonde, 31

E
Edouard, Tropical Storm, 30

F
FAA (Federal Aviation Administration), 4, 23, 42, 43
Fahrney, Delmer, 7
Fire Scout, 5
FLIR, 36
F9: The Fast Saga, 27

G
Global Hawk, 28, 29, 30, 31
GPS, 39

H
HAMSR (High Altitude MMIC Sounding Radiometer), 31
H520 (Yuneec), 21
HIWRAP (High-Altitude Imaging Wind and Rain Airborne Profiler), 31
Hurricane and Severe Storm Sentinel (HS3), 30

I
Intel, 25

J
Japan, 25

L
LIP (Lightning Instrument Packages), 31

M
Marines, US, 39
Matrice 300 RKT, 15
Matrice 200, 32, 33
Matthew, Hurricane, 31
Mavic Mini, 20
Monroe, Marilyn, 9
MQ-8 Fire Scout, 5
MQ-9 Reaper, 40, 41
MQ-1B Predator, 40, 41

N
NASA, 29, 30
National Drone Racing Championship, 13
Naval Research Laboratory, US, 7
Navy, US, 7, 8
North Atlantic Ocean, 30
Northrop Grumman, 28

O
Olympics, Summer, 25
OQ-2, 8, 9
OQ-3, 8

P
Predator, 40, 41

Q
Queen Bee, 6, 7

R
Radioplane Company, 8, 9
Radioplane One (RP-1), 8, 9
Raven, 38, 39
Reaper, 40, 41
Remote Pilot Certificate, 42, 43
RQ-11 Raven, 38, 39

S
SAR (search and rescue), 32, 33, 35
Shooting Star, 25
Sierra, 26

T
Tango II, 26, 27
TDD (Target Drone Denny), 8
Tiddledy-dee, 8
Tokyo, Japan, 25

U
United Kingdom, 6, 7
United States, 4, 7, 8, 22, 23, 42
Unmanned Aircraft General-Small (UAG), 42

W
World War II, 8

X
XM2, 26, 27

Y
Yuneec, 21